PUPPETS AND MARIONETTES

A Collector's Handbook & Price Guide

Jan Lindenberger

Schiffer Publishing Ltd

4880 Lower Valley Road, Atglen, PA 19310 USA

Antique papier mache hand puppet with real fur and wood limbs. 1940s. $60-70

Copyright © 1997 by Jan Lindenberger
Library of Congress Catalog Card Number: 97-65239

Designed by Laurie Smucker

ISBN: 0-7643-0279-5
Printed in China

Published by Schiffer Publishing, Ltd.
4880 Lower Valley Road
Atglen, PA 19310
Phone: (610) 593-1777
Fax: (610) 593-2002
E-mail:Schifferbk@aol.com
Please write for a free catalog.
This book may be purchased from the publisher.
Please include $2.95 for shipping.
Try your bookstore first.

We are interested in hearing from authors
with book ideas on related subjects.

Contents

Acknowledgments

A very special thank you to Joel Martone. Without his collection, knowledge and patience this book would not have been possible. He allowed me to bring his puppets, marionettes and ventriloquist dummies to my home and photograph his wonderful collection. Most of the puppets in this book belong to Joel and are in his shop called "Rhyme and Reason," Colorado Springs, Colorado. Many thanks, Joel.

Also thanks to Jack Underwood, from Hatfield's Antique Mall in St. Joseph Missouri, for the 1980 W.C Fields ventriloquist dummy in original box, picture; to D.L. "Butch" Ewing from Trenton, Missouri for the Three Stooges picture; and to Robin and Rennie Stein from North Miami Beach, Florida, for the pictures of several character hand puppets.

Joel Martone with (left) Bozo the clown and (right) pillow doll clown. 1970s. $100-150 each

3.5' Bozo the Clown marionette. 1970s. $100-150

Joel Martone with the marionette, Sweet Betsy from Pike, which he made by hand. Moveable eyes and mouth. 5'. 1979. $800-1000

6' plush bear holding plush 3.5' marionette bear. The large bear is made out of papier mache by Joel Martone, and covered with the fur from 28 collars. The marionette is made out of old coats. The costume fabric glows in the dark. 1980. Large bear suit $300-400. Small bear marionette. $200-250

Introduction

In the simplest terms a puppet is an inanimate figure that is made to move by human effort before an audience. But they are much more than that. Puppets and marionettes contain the magic that transforms pieces of wood, string and cloth into animated expressive characters which have captivated us all. They are whimsical and dramatic, sad and comic, but always entertaining. Since the 1940s-50s television these wooden figures have caught and kept our attention.

When the art of puppetry is at its best, it is far from our recognition. We forget that the characters are puppets, manipulated by some unseen hand. They have their own personality, carried more by their actions and comments than by their unchanging expressions, and the personality reflects the most human of emotions and thoughts, bringing us new insight and joy.

I hope the puppets and marionettes in this book awakens memories of the places they have touched and enriched your lives.

Famous Puppet Makers

Tony Sarg

Tony Sarg was a great puppeteer who started out as an illustrator in England where he was inspired by Charles Dickens. His first show was "The Old Curiosity Shop." He charged 6 pence for admission to the performance, which paid his rent. He moved to New York in 1915, where he became a puppeteer. By 1927 Tony Sarg was known as the most prolific puppeteer in America.

He performed in the 1920s Macy day parade of the balloons. His biggest show was the one he put on at the Chicago World's Fair in 1933. In 1936 President Franklin Roosevelt thought it a good idea to organize 50 marionette companies to tour the U.S.A. to explain the process of democracy and philosophy of the New Deal. Tony spearheaded this group.

Some of the greatest puppeteers studied and worked under Tony Sarg. Among them were Bill Baird who went on to a magnificent career including performing the puppet sequence in the classic movie "The Sound of Music."

Tony Sarg plastic marionette. 1930s.
$300-400

Large Tony Sarg poster, advertising his paper marionette theatre, given as a premium for A.P.W Satin Tissue. 1930. Rare. $200-250

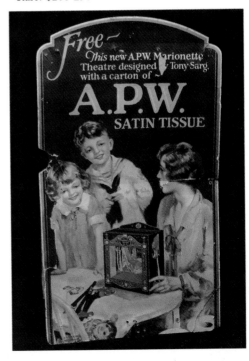

Tony Sarg A.P.W. Marionette Theatre
advertised in the poster, with cutout
puppets. 1930. Rare. $100-125

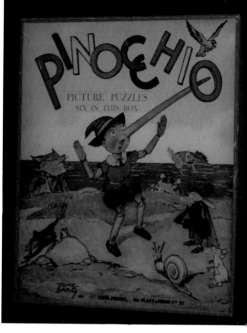

Box of six wooden Pinocchio picture
puzzles. Tony Sarg. Platt & Munk Co.,
1930s. $40-55

The Tony Sarg Marionette Book. New York:
B.W. Huebsch, 1930s. $40-50

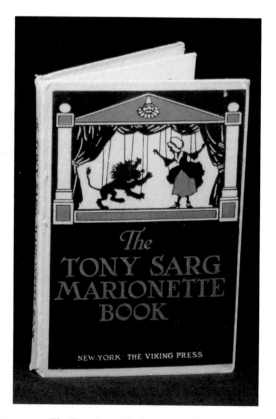

The Tony Sarg Marionette Book. New York:
Viking Press. 1940. $60-75

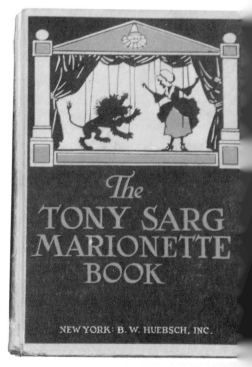

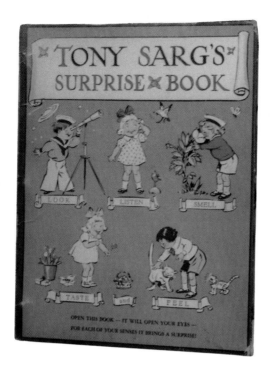

Tony Sarg's Surprise Book. Hard back.
1940. Rare. $75-100

Turnabout Theatre with Joel Martone's Handcrafted Puppets

In the 1940s the Turnabout Theatre was started in Hollywood, California, by the famous puppeteer Harry Burnette. The seating in this theatre consisted of old street car seats, which meant they spun around. The audience could watch the people in the street and then they could turn their seats around to watch the puppet show, giving the theatre its name. Some of the puppets would play out what the audience could see on the street. The original troupe were Harry Burnette, Dorothy Neuman, and Forman Brown. Forman was famous for doing many of the lines, lyrics, and music for the shows.

The puppets for the Turnabout Theatre were all hand-made and hand-carved. One of the more famous puppets was the Carman Miranda puppet. There were up to thirty puppets on one string manned by one person. The Turnabout Theatre closed and the puppets were auctioned off in 1962.

Punch and Judy

Punch and Judy is the name of the famous puppet team and is a household word throughout the English-speaking world. They performed in the world's best known puppet play.

Joel Martone

Joel Martone's interest in puppets began at the age of nine, when he cut up an old wooden orange crate with a jig saw, to make the components of a dancing marionette. His parents were very supportive of his artistic endeavors. and provided him with materials and a basement workshop. There many wonderful characters were born including a papier mache marionette, Johnny Jones, who played a smoking honky-tonk piano.

When North pole and Santa's Village opened in Cascade, Colorado, they needed a young boy to work as assistant and apprentice to their master puppeteer. Joel was given the job, and was started to work there in the summer of 1957, at the age of eleven. He learned many techniques and procedures for making and operating marionettes, and by the fall had advanced far enough to be given the job of part-time puppeteer on weekends. This soon developed into a full time job which lasted for five years, enabling him to earn enough money to pay his college tuition in California, where he majored in art.

While in college. Joel got a job building marionettes for the Macabob Toy Company in Pasadena, California. He lived an various locales in California and Utah over the course of the following fifteen years, producing many puppet shows for children and adults. He was able to purchase a number of vintage marionettes from the collection of the Turnabout Theatre in Los Angeles, California.

Mr. Martone returned to his native Colorado Springs, Colorado in 1981, where he still resides. He is co-owner of a shop there called, Rhyme and Reason, where a portion of his vast collection of puppets and marionettes are on permanent display. He enjoys showing his collection and talking about puppets to others who share his love and enthusiasm for them.

Carved wood and fabric hand puppet. 1930s. $35-45

Paper Jester puppet. 1960s. $20-30

Rod puppet, Bangkok woman. 1960s.
$90-120

Bangkok woman rod puppet. 1930s.
$150-200

Bangkok rod puppet, with flat construction.
1960s. $100-125

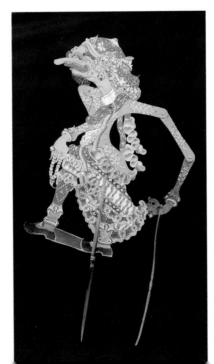

Souvenir Sicilian rod puppets. Metal bodies and suits with papier mache heads and feather head dress. 1940s. $60-100 each

Papier mache marionette from Bangkok. $50-60

Sicilian rod puppet. 36". 1960s. $65-100

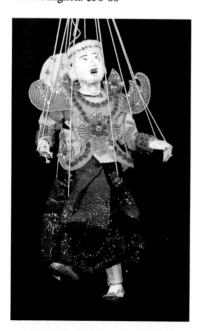 

Joel's first papier mache "Black Johnny Jones" marionette. Handcrafted wood joints. He played Tommy Dorsey's "In the Mood" on the piano. 1950s. Used at Santa's Workshop in Colorado Springs, Colorado. $200-250

Hand-carved wood old man dancing marionette. 1960s. $30-45

Cardboard puppets. Monkey and clown in original box. 1980s. $15-20 each

Cecil, Renaissance marionette by Joel. Sculpture with wood and cloth. 1980. $450-500

King marionette.
Composition,
handcrafted by Joel.
1970. $40-50

Witch from "Hansel and
Gretel," marionette.
Handmade original by
Joel. Plastic wood fiber.
1960s. $100-150

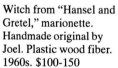

Plastic wood sculpted clown by Joel.
1970s. $30-40

Composition queen's head from a marionette by Joel. 1960s. $25-35

Santa Claus marionette dating to the 1930s, by Harry Burnett. From the Turnabout Theatre of Hollywood California. Rare. $400-450

Carman Miranda from the Turnabout Theatre of Hollywood California. 1930. $400-450

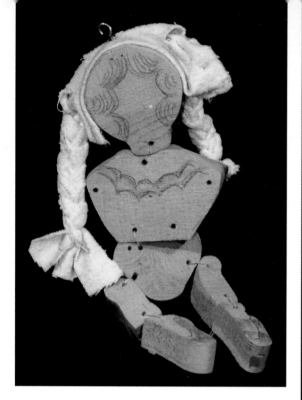

Joel's first handmade wooden puppet body, made from an orange crate. 1950s. $50-60

Rubber Policeman from "Punch and Judy" show. 1960s. $100-125

Rubber dragon hand puppet from a "Punch and Judy" show. 1970s. $70-90

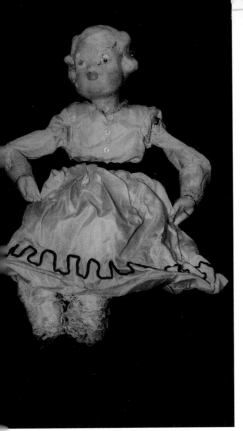

Carved composition dancing girl. 1940s. $40-50

Dwarf marionettes, dual control. Handcrafted, artist unknown. 1970s. $80-125

The Grand Turk, wood handmade composition jointed marionette. 1940. $65-80

Rubber Judy from "Punch and Judy" show. 1950s. $100-150

Dwarf marionettes with dual control. 1970s. $80-100

Original hand crafted wax Campbell Soup cook by Joel. 1960s. $75-100

Handmade marionette of old farmer with glass eyes. 1970s. $80-100

Pinocchio family, handcrafted by Joel. 1960s. $75-100 set

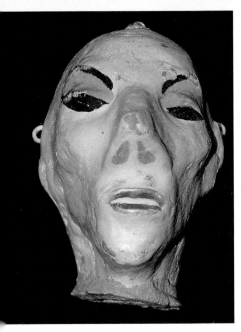

Papier mache marionette head. Handmade by Joel. 1950s. $50-60

Hand-carved marionette king. 1970s. $80-100

Buck Rogers Moon People marionette. Cloth and papier mache. 1960s. $40-50

Papier mache marionette head. 1950s. $50-60

Famous Puppet Companies

Pelham Puppets from England

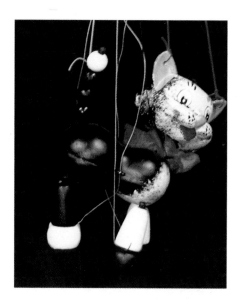

Pelham of England all wood Kitty marionette. 1970. $60-75

Wooden Pelham dragon marionette. 1960s. $80-150

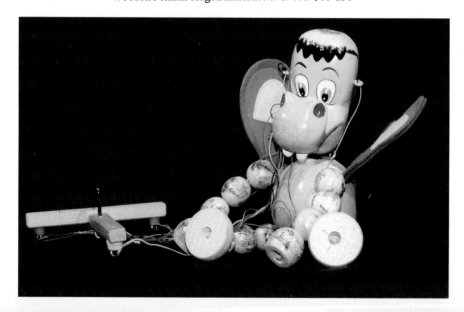

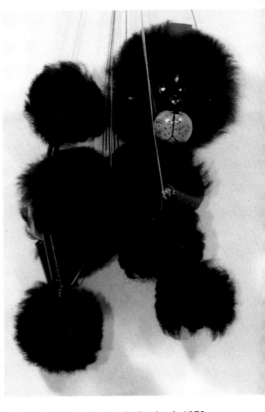

Pelham fur poodle. Made in England. 1970. $50-60

Red headed dancing boy. Pelham. 1960s. $50-60

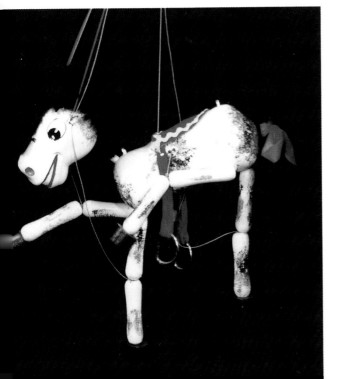

Pelham of England horse. 1970s. $75-90

Pelham wooden English girl marionette. $40-50

Pelham, composition Beatles-like character. 1970s. $75- 100

Pelham wooden marionette. 1970s. $35-45

Pelham marionette of black dancing girl. Wooden head and limbs. 1950s. $50-75

Pelham wooden cleaning lady marionette. 1970s. $60-80

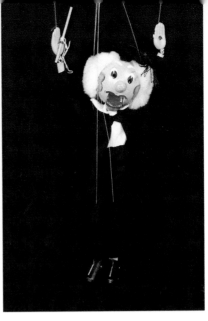

All wood conductor marionette by Pelham of England. 1960s. $65-85

Pelham teenager girl, composition and wood marionette from England. 1970s. $50-60

23

Frog marionette by Pelham. Composition.
$65-75

Hazelle Puppets

Hazelle Gypsy girl
plastic marionette.
1970s. $50- 65

Hazelle Ozark marionette. Plastic head and arms with wood body. 1950s. $45-55

Plastic Hazelle clown puppet. $80-100

Hazelle plastic Gretel marionette. 1960s. $60-75

Hazelle plastic marionette in original box. 1950s. $100-125

Hazelle's "Popular" Marionettes, black dancing man. $80- 100

Plastic sailor marionette with moveable mouth. Hazelle. 1970s. $60-75

Plastic young man marionette with moveable mouth. Hazelle. 1970s. $60-75

Plastic Scottish marionette. Hazelle. 1970s. $80- 100

Plastic Sailorette marionette with moveable mouth. Hazelle. 1970s. $60-75

Teenager marionette with vinyl head and body. 1960s. $50-65

Composition black man marionette. 1950s. $100-125

Hazelle, plastic black woman puppet. $50-60

Hazelle, plastic Robin Hood marionette. USA. 1960s. $50-75

Hazelle's plastic Hansel marionette. $60-80

Plastic "Fairy Queen" marionette. Hazelle. 1940s. $80-100

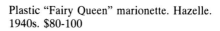

Plastic Gretel, marionette. Hazelle. 1970s. $60-80

Hazelle, plastic dancing black man marionette. 1950s. $100-150

Hazelle plastic girl marionette. 1960s. $70-80

Effanbee Puppets

Plastic head Clippo the Clown from
Effanbee. 1950s. $80-100

Clippo puppet workshop. Mint in box. Boy
and girl by Effanbee. Rare. $185-225

Rubber little girl
marionette. 1950s.
$30-40

Effanbee-Clippo the Clown puppet in original box. 1960s. $175-200

Emily Ann by EffanBee

Emily Ann marionette by Effanbee. 1940s. $100-150

Steiff Puppets

Margaret Steiff was a talented woman from Giengen, West Germany. Even though she was an invalid, she made a lasting contribution to the toy world. Her production of Steiff animals began in 1894. In 1903 Mohair was introduced and used in most of her early hand puppets. It wasn't until 1908 that glass eyes were added to the toys. And of course it isn't a Steiff bear unless it has a button in its ear.

Albert Schlopsnies worked for and designed most of the Steiff puppets and marionettes.

Steiff has four remaining factories in the world today. Two in Giengen, West Germany, one in Tunisia, and one in Austria.

Steiff products appeal to everyone: children, naturalists, artists, collectors and puppeteers. Proof positive of their identity is provided only if the famous button is in the ear.

Steiff "Merrythought Cheeky" mohair hand puppet with glass eyes and button in ear. 1960. $80-100

Steiff hand puppet, mohair brown bear with glass eyes and button in ear. 1960s. $125-150

Steiff hand puppet, mohair Lion with glass eyes and button in ear. 1950s. $75-100

Steiff hand puppet, mohair Tiger with glass eyes and button in ear. 1950. $70-100

Steiff hand puppet, mohair dog with glass eyes and button in ear. 1960. $80-100

Steiff hand puppet, mohair monkey with glass eyes and button in ear. 1950. $65-80

Gucki hand puppet from Steiff with rubber head and felt suit. 1960s. $90-125

Steiff Santa hand puppet with sculptured face and mohair beard. $60-80.

Steiff mohair squirrel hand puppet with glass eyes and Steiff button in ear. 1960. $80-100

Steiff mohair poodle hand puppet. 1955.
$75-100

Steiff hand puppet. Mohair dog with glass
eyes and button in ear. 1950. $70-90

Steiff monkey hand puppet. 1960s. $80-100

Steiff mohair fox hand
puppet. 1960s. $70-80

Steiff mohair Baby Teddy Bear hand puppet
with glass eyes. Button in ear. 1960. $65-75

Steiff bunny hand puppet. 1960s. $80-100

Steiff mohair cat puppet with glass eyes and
button in ear. 1960s. $65-80

Steiff mohair wolf hand puppet with glass eyes, shaved nose and button in ear. 1950. $75-100

Steiff mohair bulldog hand puppet with glass eyes and button in ear. 1960. $60-75

Steiff mohair dog hand puppet with large glass eyes and button in ear. 1950s. $75-100

Steiff mohair tiger hand puppet with green glass eyes and button in ear. 1960. $60-75

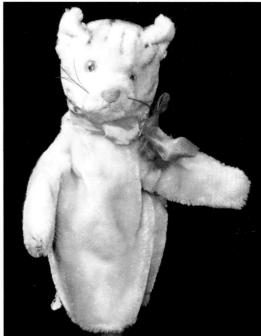

Steiff mohair dog with glass eyes and button in ear. 1960. $55-70

Steiff mohair bunny hand puppet with brown glass eyes and button in ear. 1960. $60-75

Steiff mohair alligator with glass eyes. 1960. $50-75

Steiff mohair bunny hand puppet with g eyes and button in ear. 1960s. $60-80

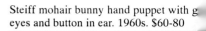

European Felt Puppets

European Felt "old lady" hand puppet with felt face. 1970s. $45-60

European Felt hand-sculptured hand puppet. 1970s. $45-60

German king, European Felt hand puppet. 1970s. $45-60

Cotton and felt jester with bead eyes. European Felt, 1970. $45-60

Pirate, felt hand puppet made in Germany.
$50-60

Steiff, Pirrot Houd, felt face hand puppet.
1960s. $25-35

European Felt king hand
puppet with felt face.
1970. $20-30

Composition head little
girl hand puppet.
European. $45-60

German girl, felt hand puppet. 1980s.
$25-30

European wooden head jester hand puppet
with felt body. 1970s. $30-40

German felt Santa hand puppet. $30-40

European Felt jester
hand puppet. 1980s.
$30-40

41

Peter Playthings Puppets

Mad Hatter composition puppet from
"Alice in Wonderland" with felt clothes.
Pelham. 1950s. $75-100

Peter Playthings monkey clown, composition. 1950s. $60-80

Composition puppet dog marionette from
"Lady and the Tramp" with felt ears and
moveable legs and head. 1960s. $100-125

Composition "Dopey" puppet head from
"Snow White and The Seven Dwarfs."
1950s. $45-60

Composition hare marionette with felt clothes. 1940s. $85-110

Peter Puppet Marionettes brochure. Peter Puppet was an early American puppet manufacturer making mostly composition puppets. $30-40

Peter Playthings composition marionette with original control. 1940s. $65-90

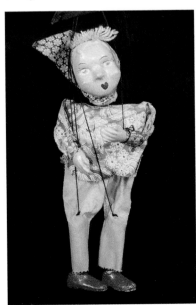

Peter Playthings composition marionette. 1940s. $60-80

Pinocchio Puppets and Collectibles

Pinocchio was born on January 25, 1883. This wooden puppet had a wish to come to life and his wish came true. It seemed every time he turned around he would get into trouble. His worst problem was that when he told a lie his nose would grow. Pinocchio had a friend that was a talking cricket named Jiminy Cricket. Gepetto was the puppets maker and his wish maker was the Blue Fairy.

In the early 1880s, an Italian journalist named C. Lorenzini needed money to pay off a gambling debt. He decided to put his hand to writing this children's book. He adopted the pen-name of Carlo Collodi wrote for a children's magazine where he gained the idea for the Pinocchio book.

His ideas for the original Pinocchio book were quite vulgar, including hangings and torture, but those ideas were soon rejected. The present story has been written in many languages and different scenes, but the true outcome is usually the same well-loved story for children of all ages.

Pinocchio hand puppet with rubber head. 1970s. $35-50

Pinocchio hand puppet with rubber head. 1960s. $30-40

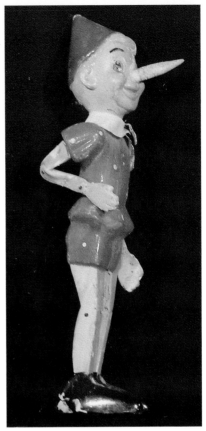

Cardboard Pinocchio theatre cut out. 1980s.
$10-15

Vinyl head Pinocchio doll with plush body.
1930s. $50-75

Pinocchio jointed rubber marionette. 1950s.
$50-75

Italian Pinocchio head,
wood. $20-30

Pinocchio vinyl head hand puppet. Gund.
1960s. $40-50

Vinyl Pinocchio doll from Knickerbocker.
1980s. $20-25

Cloth printed Pinocchio pillow doll. 1950s.
$30-40

Wooden Italian jointed
Pinocchio doll. 8".
$20-30

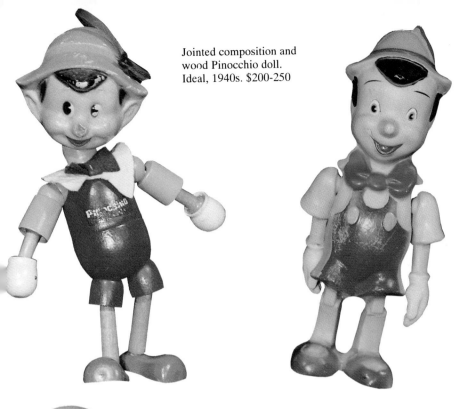

Jointed composition and
wood Pinocchio doll.
Ideal, 1940s. $200-250

All rubber jointed Pinocchio doll. 1940s.
$50-75

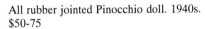

Plush Pinocchio doll. 1980s. $15-20

Jointed Pinocchio doll,
by Applause. Vinyl,
1990s. $10-15

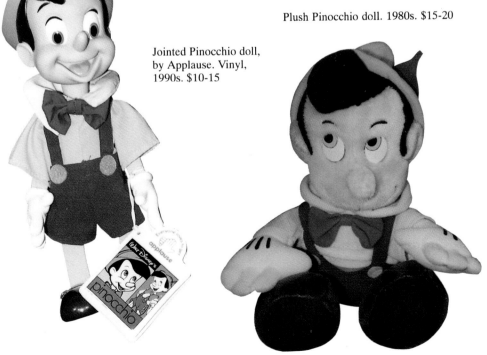

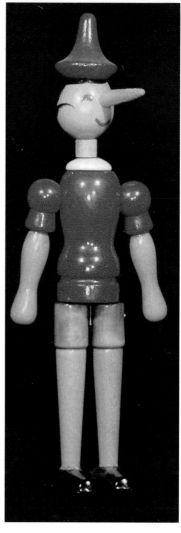

Stuffed Pinocchio doll with vinyl head and hands. 1960s. $60-75

Cotton Pinocchio pillow doll. 1970s. $15-20

Italian all wood Pinocchio doll. 1970s. $75-100

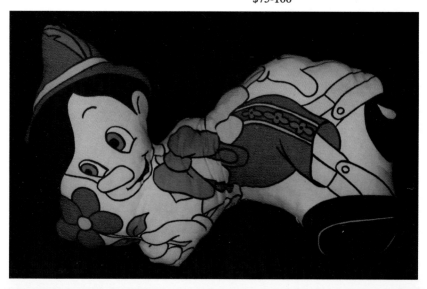

Pinocchio doll with vinyl head and plush body. Knickerbocker. 1970s. $30-40

Knickerbocker, plush stuffed Pinocchio. 1970s. $30- 40

Vinyl Pinocchio bank. 1950s. $15-20

Madam Alexander Pinocchio doll. 1980s. $75-100

Metal car with vinyl Pinocchio. $20-30

Rubber Pinocchio bank. 1970s. $30-40

Vinyl Pinocchio bank. 1975. $18-25

Vinyl unmarked Pinocchio head from a puppet. 1970s. $15-20

Jiminy Cricket hand puppet from Walt Disney Productions. Rubber head, 1960s. $35-45

Jiminy Cricket hand puppet with rubber head. Knickerbocker. 1960s. $20-25

Rubber Pinocchio bank. 1960s. $20-25

Jiminy Cricket hand puppet with rubber head. 1960s. $30-40

51

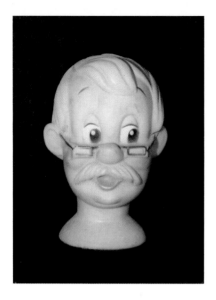

Vinyl head of Gepetto by Gund. 1970s.
$7-10

Vinyl inflatable Figaro cat. McDonalds
premium. 1990s. $6-10

Pinocchio hard back story book. 1980. $6-8

"Walt Disney's
Pinocchio" card game.
1950. $30-40

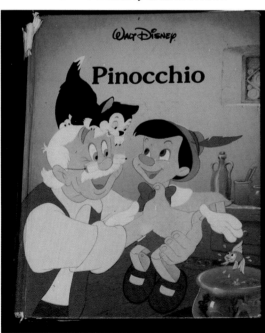

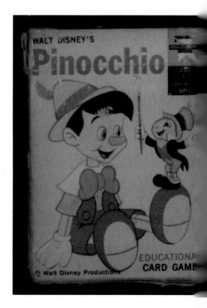

Pinocchio story book by Collodi. $15-20

Little Golden Book's *Pinocchio*. 1970. $6-8

Pinocchio, a hard back story book by Whitman. Walt Disney Productions. 1970. $10-15

Pinocchio Match 'n Color book. 1970. $12-20

Pinocchio "Giant Color/Activity Book."
Whitman Publishing Company. $6-8

Pinocchio activity book.
Games, coloring, mazes,
connect the dots and
more. 1980. $10-15

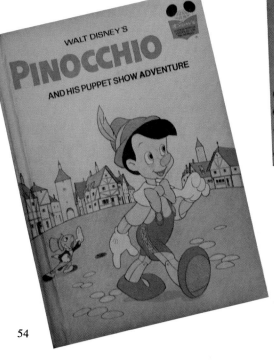

Pinocchio story book. 1975. $5-7

Disney. "Pinocchio" 1990 record. $8-10

Walt Disney's *Pinocchio,* Golden Book. 1990. $4-6

Story of Pinocchio book with audio tape. 1970s. $5-7

Little *Pinocchio* story books. 1970s. $4-6 each

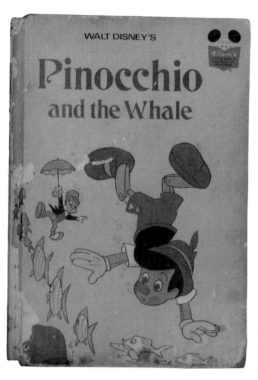

Pinocchio and the Whale story book.
$1970s. $5-7

Pinocchio Giant Fairy Story book. 1970.
$15-20

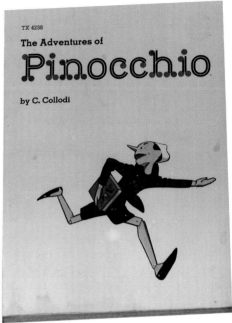

The Adventures of Pinocchio. 1960. $20-25

Pinocchio story book. 1970s. $6-10

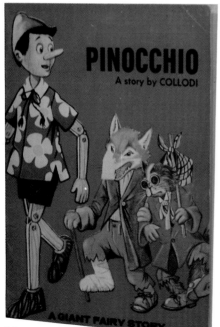

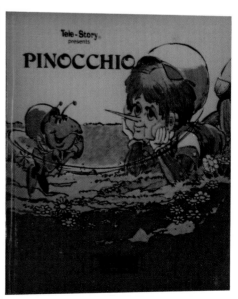

Pinocchio hard back story book. 1980.
$12-20

Walt Disney's Pinocchio "Golden Book".
1955. $20-35

Whitman Publishing, hard back *Pinocchio*
book. 1960. $12-18

Pinocchio story book and audio tape from
Walt Disney Productions. 1970s. $10-15

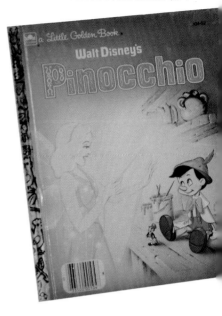

Pinocchio Golden book. 1980s. $5-7

Pinocchio Coloring Book. Whitman. 1960. $10-15

Pinocchio Plays Truant by Virginia Parsons. Hard back, 1990. $10-15

Pinocchio, a modern version of the story. Hard back. 1994. $10-15

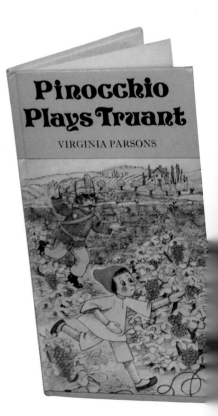

Pinocchio story book. 1978. $10-184

Walt Disney's *Pinocchio: Fun with Shapes & Sizes,* story book. 1990. $6-10

Adventures of Pinocchio by Carlo Collodi. Hard back. 1940. $20-30

The Wonderful Adventures of Pinocchio comic book. 1950. Originally 12 cents. $30-40

Golden Sound Story of *Pinocchio.* 1990.
$10-15

Classic Junior *Pinocchio* comic. $20-30

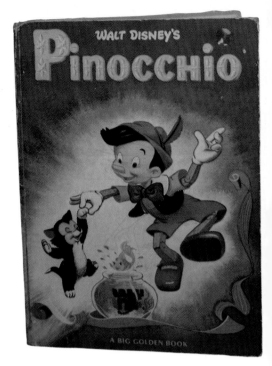

Golden Book, Walt Disney's *Pinocchio.*
1960. $20-30

Big Little Book of *Pinocchio and Jiminy
Cricket.* 1940. $30-40

Disney's *Pinocchio* story book. Marigold Press. 1939. $65-80

Walt Disney's version of Pinocchio story book. 1939. $60-75

Make a Pinocchio String Puppet. $10-15

Cardboard puzzle. 1960. $10-15

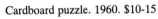

"Pinocchio Game." Walt Disney Productions. 1970s. $30-40

Cotton Pinocchio twin size sheet. 1980s. $15-20

Jimmy Nelson and Danny O'Day "Pinocchio" record. 1960s. $30-40

Pinocchio child's vinyl purse. 1990s. $8-10

Plastic Pinocchio plate. "Lead me to Devil
Park." $8- 10

Child's silver Pinocchio spoon. 1950s.
$30-45

Howdy Doody Puppets and Collectibles

The nationwide love affair between Buffalo Bob, Howdy Doody, and the Peanut Gallery first took shape December 27, 1947.

For 13 years and 2,544 shows Buffalo Bob and Howdy Doody were an American institution on NBC television. The show closed down in September, 1960.

The real Howdy Doody resides in a glass case in Buffalo Bob's North Carolina home. The freckled face boy, dressed in western attire, was a composite creation from several designs by Walt Disney artists.

Once the look was established, Velma Dawson, a sculptress from California, created the puppet.

Bob Keeshan, a page and errand boy on the show, became the mute Clarabell the clown. This horn-honking seltzer-squirting, prankster would create havoc daily on the set. In time Bob Keeshan left the show and became the famous character Captain Kangaroo.

Rhonda Mann was the voice of Princess Summer Fall Winter Spring, of the Tinka Tonka Tribe. Howdy's voice was that of Buffalo Bob.

Some Howdy's notable friends included Dilly Dally, the troublesome, and the cantankerous Phineas T. Bluster. Flub A Dub was a mythical creature made up of eight different animals: a cat's whiskers, cocker spaniel's ears. a giraffe's neck, duck's bill, dachshund's body, a pig's tail, seal's flippers and the memory of an elephant.

Howdy Doody hand puppet. Rubber head.
1950s-60s. $50-60

Clarabell hand puppet with rubber head.
1950s. $20-30

Howdy Doody hand puppet with vinyl head.
(Note: suit has pattern from the show
printed on it.) 1950s. $50-60

Howdy Doody hand puppet with rubber
head. 1950s. $50-60

Clarabell hand puppet vinyl head. 1950s-60s. $35-45

Howdy Doody hand puppet in western outfit. 1960s. $40-50

Terry cloth Howdy Doody hand puppet. 1950s. $15-20

Terry cloth Howdy Doody hand puppet/ wash cloth. 1950s. $25-35

Howdy Doody ventriloquist dummy. 1950s.
$250-350

Composition ventriloquist dummy,
Howdy Doody. 1950s. $200-250

Princess Summer Fall Winter Spring plastic
marionette with moveable mouth. Hazelle.
1980s. $65-80

Dilly Dally hand puppet with rubber head.
1950s. $15- 20

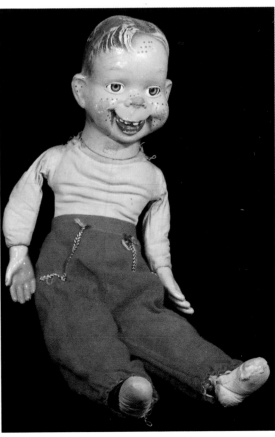

Princess Summer Fall Winter Spring. 1970s. $30-40

Rubber head Howdy Doody hand puppet. 1950s. $30-40

Dilly Dally hand puppet, rubber head. 1960s. $50-60

Plastic moveable mouth Dilly Dally from "Howdy Doody." 1950s. $20-30

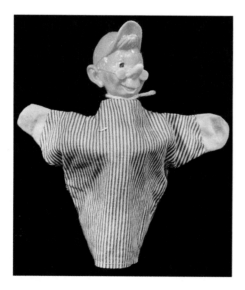

Howdy Doody dummy in original outfit.
1970s. $100- 150

Plastic Howdy Doody figure that sits
on a record as it spins. 1950s. $40-50

Dilly Dally hand puppet. Rubber head.
$30-40

Phineas T. Bluster hand puppet with rubber
head, from "Howdy Doody Show."
1950s-60s. $35-40

Phineas T. Bluster hand puppet with rubber head. 1950s-60s. $35-40

Phineas T. Bluster hand puppet. Rubber head, 1960s. $35-40

Dilly Dally composition marionette from "Howdy Doody Show." Peter Plaything. 1950s. $100-125

Flub a Dub hand puppet bird from "Howdy Doody Show." Vinyl head. 1960s. $40-50

Dilly Dally hand puppet, with vinyl head. 1950s-60s. $40-50

Glass Howdy Doody night light. 1960s. $40-50

Flub a Dub marionette from "Howdy Doody." 1950s. $80- 100

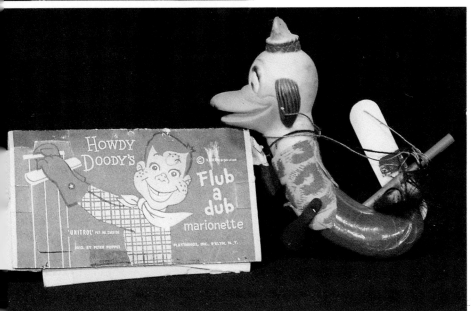

Bean bag Howdy Doody doll with rubber
head. 1980s. $15-20

Cloth Howdy Doody doll with printed face.
8". 1990. $25-35

Cloth Howdy Doody doll with printed face.
12". 1990. $25-35

Clarabell doll from the "Howdy Doody
Show." 1970s. $50-75

Cloth Howdy Doody doll from Applause. 1980s. $45-60

Howdy Doody doll. Vinyl. 1980s. $50-65

Princess Summer Fall Winter Spring hand puppet, from the "Howdy Doody Show." 1965. $40-50

Plastic Howdy Doody plate. 1950s. $40-50

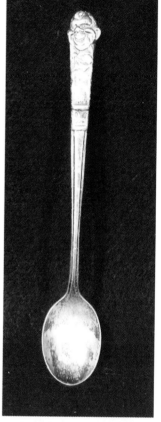

Child's silver Howdy Doody spoon. 1950s. $30-45

Howdy Doody punch cards. 1950s. $15-20

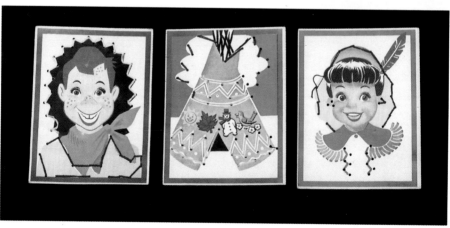

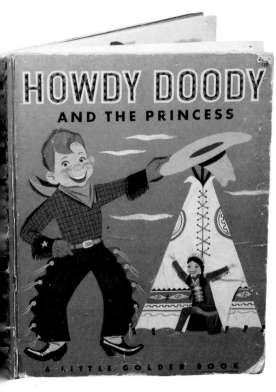

Howdy Doody magazine ad for Kellogg's
Rice Krispies. 1950s. $15-20

Golden Book, *Howdy Doody and the
Princess.* 1952. $45-55

*Say kids! What Time Is It? Notes from the
Peanut Gallery.* Stephen Davis. Hard back.
1980's. $10-15

Howdy Doody cardboard giant puzzle.
1950. $25-35

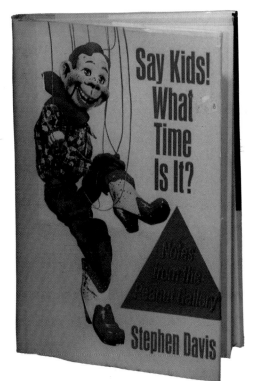

"Howdy Doody and the Air-O-Doodle."
Record, RCA. 1950s. $40-50

"Howdy Doody Magic Kit." A premium
from Luden's, Inc. 1950s. Rare. $30-40

Life magazine photo of Buffalo Bill and his
collection. 1950s. $15-20

Life magazine article accompanying the
photo, shows the original Howdy Doody.
1960. $15-20

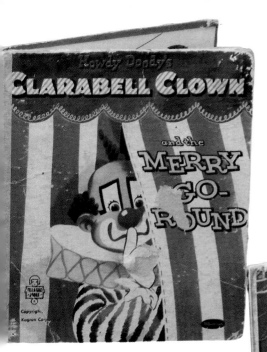

Clarabell Clown and the Merry-Go-Round,
Tell a Tale book. 1960s. $30-35

Golden Book, *It's Howdy Doody Time.*
1952. $35-40

Golden Book, *Howdy Doody and Mr.
Bluster.* 1952. $35-40

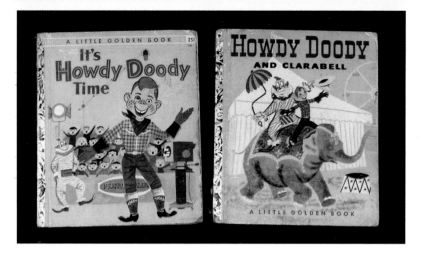

Howdy Doody story books from Golden
Book. 1950s. $12-15 each

Howdy Doody Ice Milk Bars advertising
label. 1950s. $30-40

Original prints from the "Howdy
Doody Show." 1950s. $40-50 each

Ventriloquists Dummies and Marionettes

Charlie McCarthy
dummy, 1960s.
$200-300

Jerry Mahoney
ventriloquist dummy.
1960s. $100-150

Composition Charlie McCarthy dummy. 1940s. $575-700

Charlie McCarthy vinyl dummy in original clothes. 1970s. $100-150

Bakelite Charlie McCarthy radio. 1940s.

W.C. Fields ventriloquist dummy in
original box. Celebrity. 30" Juno
Goldberger doll Mfg. EEGEEE Company.
Vinyl. 1980s. $200-300

Jerry Mahoney ventriloquist dummy,
composition head and limbs. 1940.
$650-800

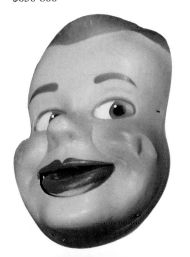

Plastic face of Jerry Mahoney with
moveable mouth. 1960. $30-40

Plastic head and limbs ventriloquist dummy with original clothes. 1950s. $75-100

Jerry Mahoney, plastic, dummy. 1970s. $75-100

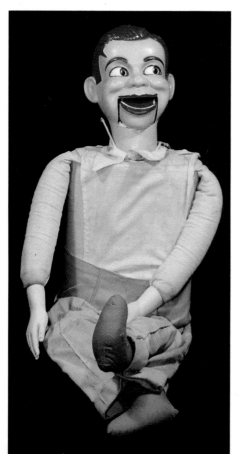

Danny O'Day vinyl dummy. 1980s. $60-75

Dummy Dan, The Ventriloquist Man. A take-off on Charlie McCarthy, this composition doll is in its original box. 1920s. $100-125

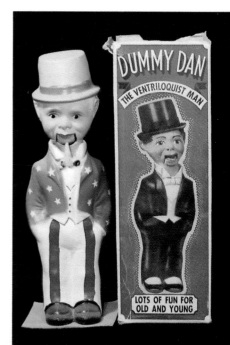

Danny O'Day vinyl ventriloquist. 1960s.
$100-125

Emmett Kelly ventriloquist dummy, vinyl, dummy. 1980. $75- 100

Santa puppet with Jerry Mahoney moveable head. $40- 50

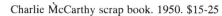
Charlie McCarthy scrap book. 1950. $15-25

Vinyl Mickey Mouse dummy. 1980. $60-75

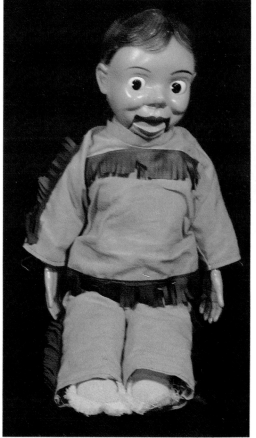

Rubber Davy Crockett dummy. 1950s.
$100-125

Danny O'Day dummy, composition. 1970s. $60-75

Simon Sez dummy. Vinyl head and limbs. 1970s. $100- 135

Plastic ventriloquist dummy. Teenage girl. 1980. $30-40

Marionettes

Composition marionette of black man. Jointed arms and legs with moveable mechanical nodding head. 1920s. $3,500-4,500

Black minstrel marionette. 1880s-90s. $350-450

Brown minstrel marionette. 1880s-90s.
$350-450

Vinyl Popeye marionette by Knickerbocker.
1950s. $40-50

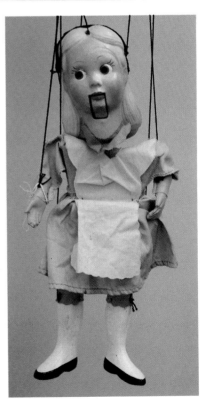

Alice from "Alice in Wonderland."
Composition. 1940s-50s. $175-250

Push button Raggedy Ann marionette by
Knickerbocker. Original box. 1980s. $40-50

Composition and vinyl clown puppet.
Mexico, 1980s. $15-20

Plastic Charlie Brown marionette. Pelham
of England, 1979. $85-125

All rubber Monkey marionette. 1950s.
$35-45

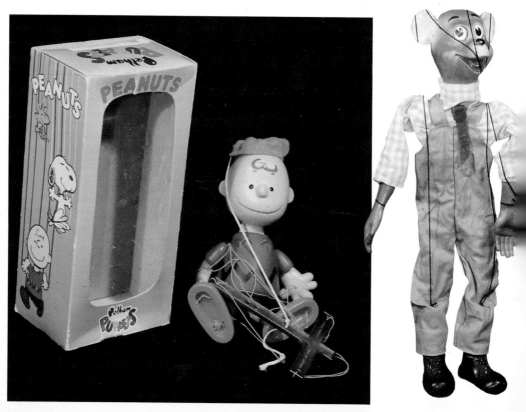

Wooden marionette horse from Iran. 1970s. $35-45

Rubber European lady marionette. 1960s. $50-60

All rubber girl marionette. 1960s. $60-75

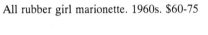

Plastic and vinyl Japanese man marionette. 1960s. $20-25

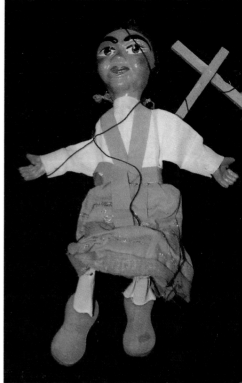

Black marionette with composition head.
1970s. $40-50

Hand-carved wooden marionette of a lady
with moveable mouth and jointed limbs.
1970s. $50-65

Mexican child marionette. Composition
head and hands. 1970s. $10-15

Pelham bear marionette with composition
parts. 1960s. $45-55

Hand carved Hillbilly woman marionette. 1960s. $60- 80

Composition Mexican marionette. 1980s. $15-20

Old Mexican marionette. 1970s. $20-30

King marionette with wire wrapped hands and jointed body. 1960s. $60-75

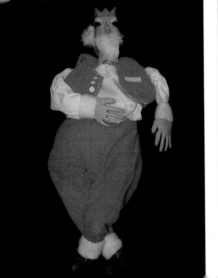

Spanish señorita composition marionette.
1950s. $40- 60

Little Red Riding Hood, composition
marionette. Hand crafted by Joel. 1960s.
$80-100

Captain Hook
marionette from
"Peter Pan."
Composition.
1960s. $80-100

Plastic clown marionette from Hazelle.
1960s. $40-50

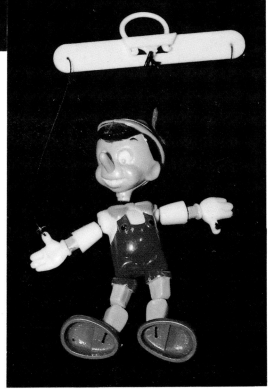

Plastic jointed Pinocchio marionette.
1970s. $40-50

Famous Television and Movie Character Hand Puppets

Shari Lewis

Shari Lewis was a much loved ventriloquist from the 1950s. In 1953 her father brought her to his T.V. studio, where he gave her a pen and paper and told her to write her own show, which she did. He then took the idea to Sid Caesar who approved it. The Shari Lewis Show was replaced in 1963 by The Chipmunks. The star of her show was the delightful, kindhearted, Lambchop. Shari and her puppets are again doing the Shari Lewis show.

Jerry Mahoney felt body hand puppet with composition head. 1940s. $60-80

Charlie McCarthy hand puppet with composition head. 1940s. $100-125

Moe, Larry, and Curly, "The Three
Stooges" hand puppet set, with rubber
heads and cloth suits. 1940s. $200-250 set

The Three Stooges finger puppets with
rubber heads. 194s. $50-60 each

Larry hand puppet with rubber head. 1940s.
$60-75

Laurel hand puppet with rubber head.
Knickerbocker. 1960s. $35-45

Hardy hand puppet with rubber head.
Knickerbocker. 1960s. $35-45

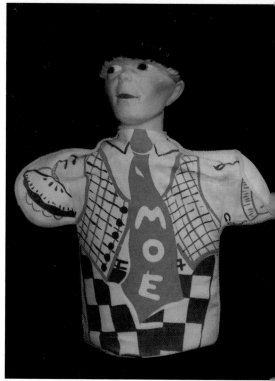

Moe hand puppet with rubber head. 40s.
$60-75

Captain Kangaroo rubber head hand
puppet. 1950s. $30-40

Wimpy hand puppet by Gund. Rubber head. 1970s. $30-40

Rubber head Brutus, Popeye's friend, hand puppet. 1970s. $20-30

Vinyl head of a "Wimpy" puppet from the "Popeye" gang. 1970. $20-30

Popeye rubber head hand puppet. 1970s. $15-20

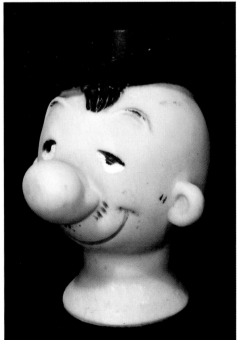

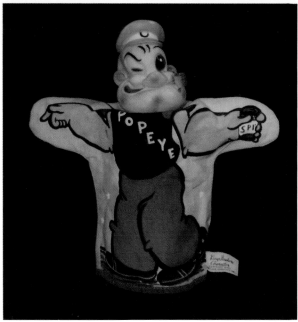

Popeye hand puppet with rubber head.
Knickerbocker. 1970s. $20-25

Olive Oyl hand puppet. 1970s. $18-25

Olive Oyl hand puppet with rubber head.
1970s. $18-25

Vinyl head Olive Oyl puppet.
Knickerbocker, King Features. $20-30

Sweet Pea vinyl head hand puppet. 1980s. $10-15

Rubber head Dopey hand puppet from "Snow White and the Seven Dwarfs." 1970s. $25-30

Sweet Pea hand puppet with rubber head. 1970s. $18-25

Rubber head Dopey hand puppet. 1960s.
$20-25

Sleeping Beauty, rubber head hand puppet.
1970s. $20- 30

Dopey plush and felt hand puppet. 1990s.
$7-10

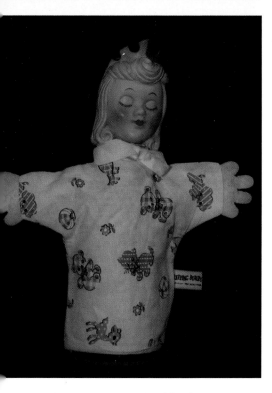

Sleeping Beauty, rubber head hand puppet.
1960s. $20-30

Merry Good Witch from "Sleeping Beauty,"
rubber head hand puppet. Knickerbocker.
1960s. $20-30

Prince Charming from "Sleeping Beauty."
Rubber head hand puppet. Gund. 1970s.
$25-35

Witch from "Sleeping Beauty." Rubber
puppet head. 1970s. $15-20

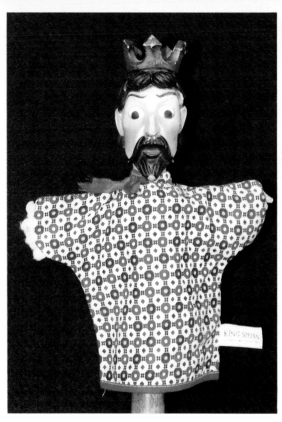

Rubber head Tinkerbell. Walt Disney
Productions. 1970s. $30-40

King from "Sleeping Beauty," vinyl head
hand puppet. 1970s. $20-25

Rubber head Peter Pan hand puppet. Gund.
1960s. $20- 30

Tinker Bell, rubber head hand puppet.
Knickerbocker. 1950s. $20-25

Wendy from "Peter Pan," rubber head hand
puppet. Ideal. $30-35

Cotton Sorcerer hand puppet with rubber
head from "Sleeping Beauty." 1970s.
$30-40

Rubber Ring Master hand puppet. 1960s.
$50-65

Humpty Dumpty hand body with molded plastic face and cloth body. 1970s. $20-30

Rubber head Scarecrow from "The Wizard of Oz." 1960s. $15-20

Rubber head Mr. Macabob hand puppet from Macabob Toy Co. Pasadena, California. 1960s. $30-40

Mary Poppins rubber head, hand puppet.
1950s. $30- 40

Rubber head Zorro hand puppet. 1970s.
$20-30

Queen Mother, rubber head hand puppet
from "Alice in Wonderland." 1970s. $20-25

Batman rubber head, hand puppet. 1950s.
$30-40

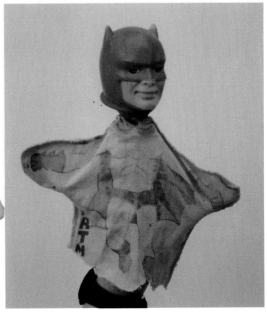

Dennis the Menace rubber
head hand puppet. 1960s.
$10-15

Rubber head Superman hand puppet.
1970s. $30-40

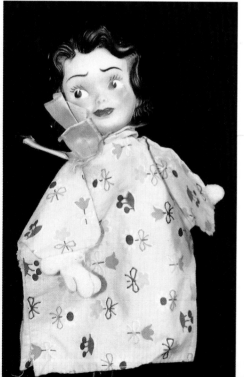

Rubber head Lois Lane hand puppet.
European-Gund. 1965. $20-30

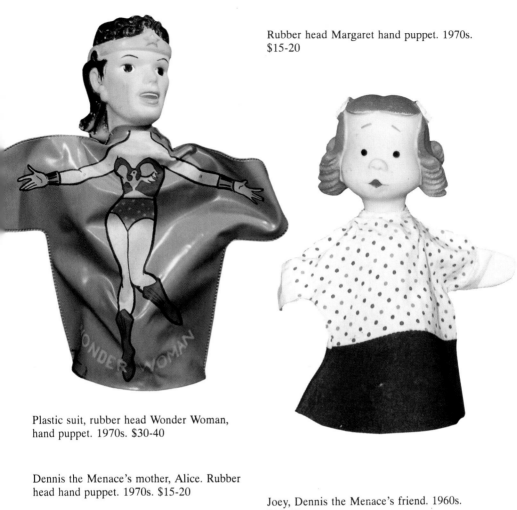

Rubber head Margaret hand puppet. 1970s.
$15-20

Plastic suit, rubber head Wonder Woman,
hand puppet. 1970s. $30-40

Dennis the Menace's mother, Alice. Rubber
head hand puppet. 1970s. $15-20

Joey, Dennis the Menace's friend. 1960s.
$15-20

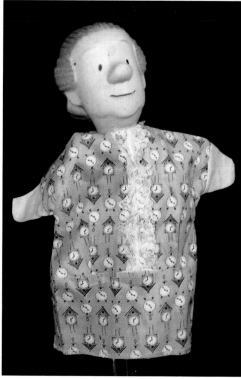

Felt Mr. Wilson hand puppet from "Dennis the Menace." 1970s. $15-20

Alice, Dennis the Menace's mother. Rubber head hand puppet. 1970s. $15-20

Mrs. Wilson, Dennis the Menace's neighbor. Rubber head hand puppet. 1970s. $15-20

Dennis the Menace's father. Rubber head hand puppet. 1970s. $15-20

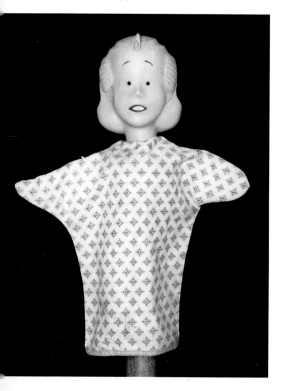

Dennis the Menace rubber head hand puppet. 1970s. $15-20

Vinyl head, Archie, hand puppet. 1970s. $20-30

Betty from the "Archie" gang. Rubber head, hand puppet. 1960s. $20-30

Vinyl head Fred Flintstone hand puppet. 1970s. $30-40

Barney Rubble from "The Flintstones," rubber head hand puppet. 1970s. $15-20

Bonnie Braids rubber head hand puppet. $40-45

Fred Flintstone, rubber head hand puppet. 1970s. $15-20

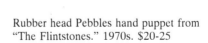

Rubber head Pebbles hand puppet from "The Flintstones." 1970s. $20-25

Betty Rubble from "The Flintstones," rubber head hand puppet. 1970s. $15-20

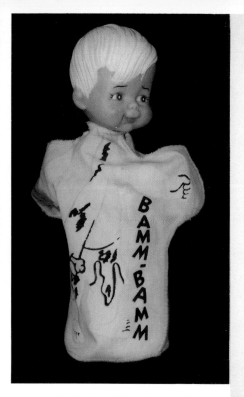

Rubber head Bamm Bamm from
"The Flintstones." 1970s. $15-20

Grandpa, rubber head hand puppet from
"The Munsters." 1970s. $20-30

Herman Munster rubber hand puppet, from
"The Munsters" television show. 1970s.
$125-150

Morticia from "The Addams Family" show,
rubber head hand puppet. 1970s. $20-30

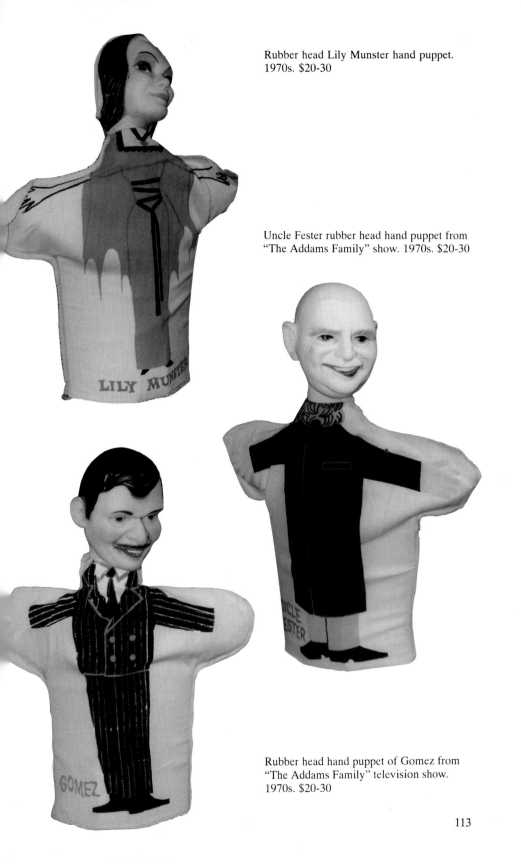

Rubber head Lily Munster hand puppet. 1970s. $20-30

Uncle Fester rubber head hand puppet from "The Addams Family" show. 1970s. $20-30

Rubber head hand puppet of Gomez from "The Addams Family" television show. 1970s. $20-30

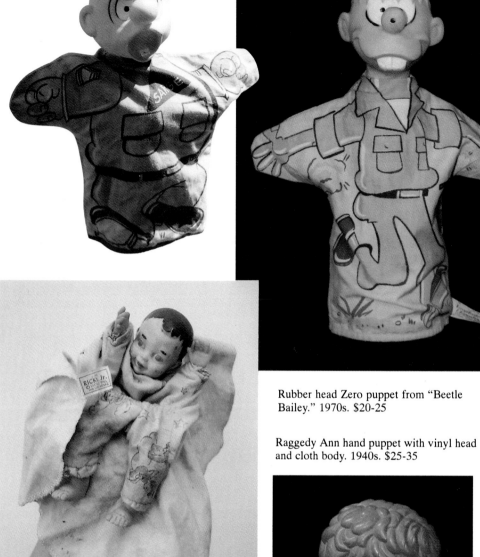

Rubber head Zero puppet from "Beetle Bailey." 1970s. $20-25

Raggedy Ann hand puppet with vinyl head and cloth body. 1940s. $25-35

Top photo:
Sarge from "Beetle Bailey," rubber head hand puppet. 1960s. $20-30

Baby Ricki Jr. rubber hand puppet from the "I Love Lucy" show. 1960s. $45-60

Vinyl head Raggedy Ann hand puppet.
1960s. $25-30

Dr. Doolittle hand talking puppet. 1970s.
$40-50

Monkees singer, Mickey Dolenz, hand
puppet with rubber head and plastic suit.
1967. $25-35

Scar, finger puppet, Burger King premium from "The Lion King." 1994. $6-8

The Monkees, four-in-one hand puppet. Rubber heads. 1960s. $200-300

Rubber President Clinton, hand puppet. Hand Critters. 1980s. $15-20

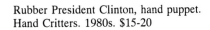

Bozo the Clown, rubber head hand puppet. 1970s. $15- 20

Bob Hope hand puppet with vinyl head and cloth body . 1970's. $40-55

Jerry Lewis rubber head hand puppet. 1960. $15-20

Wooden head soldier from "Babes in Toyland." 1970s. $15-20

Hamburger police from McDonalds, rubber head hand puppet. $20-30

Gargoyle hand puppet vinyl hand puppet from Disney's "Hunchback of Notre Dame." 1994. $7-10

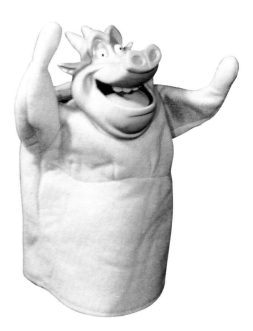

"Pocahontas" finger puppet from fast food restaurant. 1990s. $3-5

Rubber teaching hand puppets. 1980s.
$15-20 each

Boar finger puppet, a McDonald's premium
from "The Lion King." 1980s. $10-15

Rubber teaching hand puppet. 1980s.
$15-20

Plush Pocahontas dog hand puppet. 1994.
$10-15

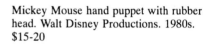

Plush Minnie Mouse hand puppet. 1980s. $8-10

Rubber head Mickey Mouse hand puppet.
Gund. 1965. $20-25

Mickey Mouse hand puppet with rubber head. Walt Disney Productions. 1980s. $15-20

Mickey Mouse rubber head hand puppet.
Walt Disney. 1970s. $20-30

Rubber head Minnie Mouse hand puppet
from Knickerbocker. 1970s. $25-35

Minnie Mouse with rubber head and cloth
body hand puppet from Walt Disney
Productions. 1970s. $20-30

Plush Mickey Mouse hand puppet. 1990s.
$16-20

Rubber head Donald Duck, hand puppet.
1950s. $60-70

Plush Tweety bird hand puppet. 1985.
$8-12

Rubber Porky Pig hand puppet. 1950s.
$50-60

Pluto hand puppet with rubber head and
cloth printed body. Walt Disney Productions. 1970s. $20-30

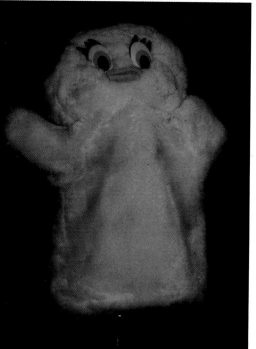

Rubber Donald Duck puppet Walt Disney Productions. 1940s. $50-60

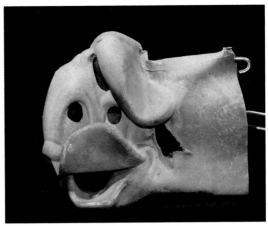

Plush Daffy Duck hand puppet by Warner Brothers. 1980s. $10-15

Plush Bugs Bunny hand puppet. 1980s. Warner Bros. $8-12

Walt Disney's Donald Duck hand puppet with rubber head. 1970s. $30-40

Jerry from "The Tom and Jerry Show," with rubber head and plush body. 1960s. $20-30

One of "The Three Little Pigs," rubber head hand puppet. 1950s. $30-40

Rubber head Goofy hand puppet. Walt Disney. Gund. 1970s. $15-20

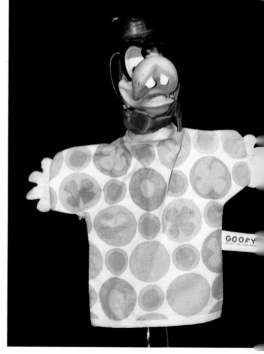

Rubber head pig from "The Three Little Pigs," hand puppet. 1950s. $40-50

Tom from "The Tom and Jerry Show," hand puppet with pull string for voice. 1960s. $20-30

Plush Sylvester the Cat hand puppet. 1990s. $9-14

Another hand puppet from "The Three Little Pigs." 1960s. $30-40

Mr. Magoo rubber head hand puppet. 1970s. $20-30

Rubber head cat hand puppet with googly eyes. 1950s. $30-40

Rubber head Chipmunk puppet from "Chip and Dale" show. 1950s-60s. $30-45

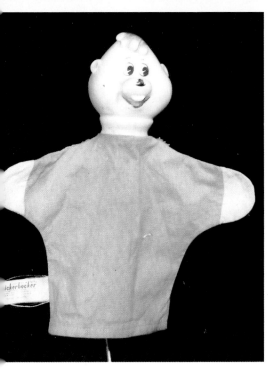

Rubber head Alvin from "Alvin and the Chipmunks." Knickerbocker. 1970s. $20-30

Alvin the Chipmunk. Vinyl head hand puppet. 1970. $20-30

Alvin from "Alvin and the Chipmunks," plush hand puppet. 1980s. $15-20

Huckleberry Hound, rubber head hand puppet. 1970s. $20-30

Dutch Boy Paint advertising hand puppet. 1950s. $30- 40

Rubber head Campbell Soup kid. 1950s. $40-50

Woody hand puppet. A McDonald's premium from Disney's "Toy Story." 1996. $10-15

Toy Works Buzz Lightyear hand puppet, McDonalds premium. 1990s. $10-15

Burger King premium, dog hand puppet
from "Pocahontas" movie. 1990s. $8-12

Pumbaa and Rafiki finger puppets from
"The Lion King" movie. Premiums from
Burger King. 1990s. $3-5 each

Finger puppets from "The Lion King"
movie. Premiums from Burger King. 1990s.
$5-6 each

Bullwinkle, rubber head hand puppet. 1950s. $35-45

Dr. Seuss's "Cat in the Hat." Rubber head and plush body puppet. 1960s. $20-30

Rubber head, Pedro, from "Lady and the Tramp." 1965. $20-25

Rooster from a chicken food advertisement, rubber head hand puppet. 1970s. $15-20

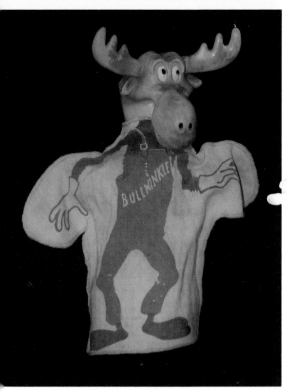

Rubber head Bullwinkle hand puppet. 1970s. $20-25

Rubber head Baby Huey hand puppet. Knickerbocker. 1950s. $20-30

Rubber head Woody the Woodpecker hand puppet. 1960s. $20-30

Furry Shaggy Dog hand puppet. 1990s. $5-8

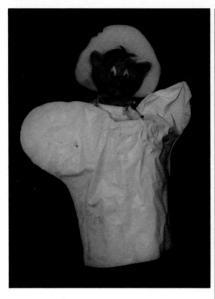

Speedy Gonzalez hand puppet, rubber
head and plastic suit. 1965. $15-20

Rubber head Archimedes hand puppet.
1970s. $20-25

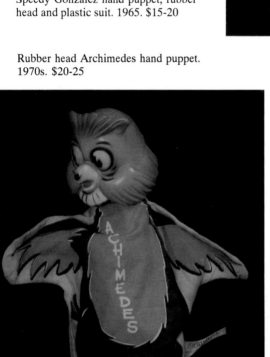

Rubber head Rocky the Squirrel hand
puppet. 1960s. $20-30

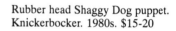

Rubber head Shaggy Dog puppet.
Knickerbocker. 1980s. $15-20

Rubber head Heckel and Jeckel crows from
an early cartoon show. 1940s-50s. $40-50
each

Dog from "Lady and the Tramp." Gund.
1970s. $20-30

Tramp, from "Lady and the Tramp" hand
puppet, rubber head. 1960s. $20-30

Chip the monkey. Rubber head hand puppet. 1965. $ 25-35

Rubber head Yogi the Bear hand puppet. 1970s. $15-20

Rubber head Pongo from "101 Dalmatians," hand puppet. 1980s. $20-30

Yogi the Bear velour hand puppet. 1970s. $10-15

Baby Huey vinyl hand puppet.
Knickerbocker. 1960s. $35-45

Clarabell Cow rubber head hand puppet.
1960s. $30-40

Charlie Horse rubber head hand puppet. $30-40

Lamb Chop from the "Shari Lewis Show." Rubber head hand puppet. 1970s. $20-30

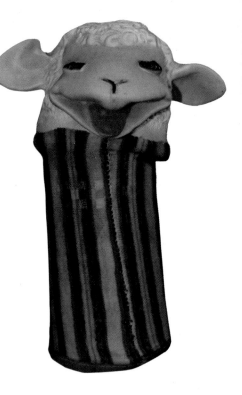

Charlie Horse from the "Shari Lewis Show." Fabric and wire. 1970s. $25-35

Plush Christmas Lamb Chop hand puppet from "Shari Lewis Show." 1980s. $10-15

Charlie Horse cotton hand puppet from the "Shari Lewis Show." 1990. $25-30

Plush Baby Lamb Chop with pacifier. Hand puppet. 1990s. $10-15

Rubber head Hush Puppy hand puppet, from Shari Lewis. 1970s. $10-15

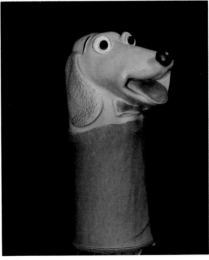

Jim Hensen's Muppet Bert, hand puppet. 1980. $10-15

Rubber Yoda puppet. Lucas Films. 1981. $60-80

Little Sprout, Green Giant hand puppet. 1980s. $10- 15

Plush Big Bird hand puppet. 1980s. $10-15

Plush Gremlin from the "Gremlins" movie.
1980s. $40-50

Hush Puppy from the "Shari Lewis Show."
1970. $20-30

Ernie felt hand puppet. 1980s. $15-20

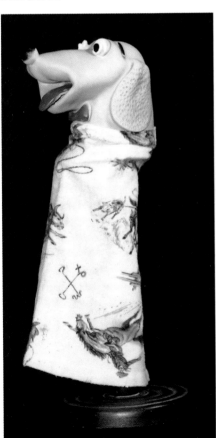

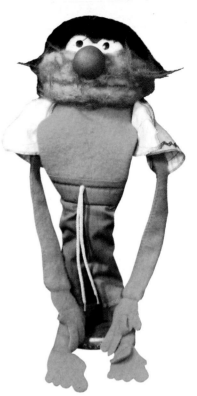

Kermit the Frog by Jim Hensen, hand puppet. 1980s. $20-30

Miss Piggy vinyl hand puppet. Knickerbocker. 1970s. $25-30

Jim Hensen's, Muppet character, "Animal," hand puppet with extended body. $20-30

Velour Kermit the Frog hand puppet. 1980s. $10-15

Berenstain Bear Dr. Gert Grizzly, cotton hand puppet. 1980s. $15-20

Plush PaPa Smurf hand puppet. 1980s. $15-20

Cookie Monster and Big Bird glove puppets. 1990s. $4-5 each

Plush dog, Rolff, hand puppet from Jim Hensen's Muppets. 1980s. $15-20

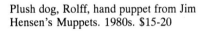

Fabric Berenstain Bears hand puppets,
Grizzly Gramps and Sister Bear. 1985.
$10-15 each

Gumby's Pokey horse hand puppet.
Rubber head. 1970 $20-25

Plush Snuggles hand puppet. Advertising
for Downy Fabric Softener. 1990s. $15-20

Plush Smokey the Bear, hand puppet.
1980s. $15-20

Topo Gigio from Lamar Toy Co. Rubber
head. 1964. $40-50

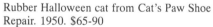

Plush Garfield hand puppet. 1990s. $10-15

Rubber Halloween cat from Cat's Paw Shoe
Repair. 1950. $65-90

Miscellaneous Hand Puppets

Plush and cotton

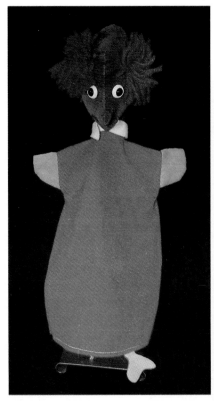

Wood and mohair Church hand puppet. 1980s. $8-12

Wood and mohair Church hand puppet. 1980s. $8-12

Felt handmade fairy, hand puppet. 1970s. $8-12

Plush dog hand puppet. 1970s. $8-12

Plush fantasy dragon hand puppet. Japan. 1982. $9-15

Plush Donkey hand puppet. 1990. $8-12

Plush Rooster hand puppet. 1980s. $7-10

Plush toucan hand puppet with button eyes. 1990. $10- 15

Plush raccoon hand puppet. 1990s. $7-10

Freddie the Frog hand puppet. Plush head, cloth body. 1980s. $8-14

Plush velour robins in a nest, finger puppets. Commercial. 1990s. $7-10

Velour and plush hand puppet. Guard dog for safety. 1980s. $15-20

Plush bear hand puppet with bead eyes. 1970s. $25- 40

Plush muskrat hand puppet. Fairy Tale Division. $10-15

Plush floppy puppet. 1990s. $10-15

Plush kangaroo hand puppet with baby in pocket. 1980. $10-15

Commercial Dragon fly Lady. Cotton and felt. 1980s. $8-15

Plush Mo Mo cow hand puppet. $6-10

Plush wolf hand puppet. "M" mascot. 1980s. $10-15

Plush hippo, hand puppet. 1990s. $8-15

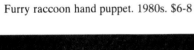
Furry raccoon hand puppet. 1980s. $6-8

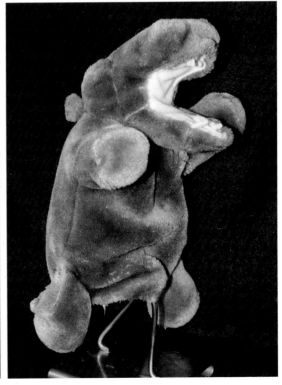

Plush bunny hand puppet. 1980s. $8-12

Plush lion hand puppet. 1990s. $10-15

Plush unicorn hand puppet from Gund. 1980s. $8-12

Plush reindeer hand puppet with button eyes. 1980s. $7-10

Cotton pig hand puppet. McDonald's premium, 1990s. $8-15

Plush American Character rabbit hand puppet. 1970s. $30-40

Plush monkey hand puppet. Glass eyes. American Character, 1970s. $30-40

Santa hand puppet, felt. 1975. $15-20

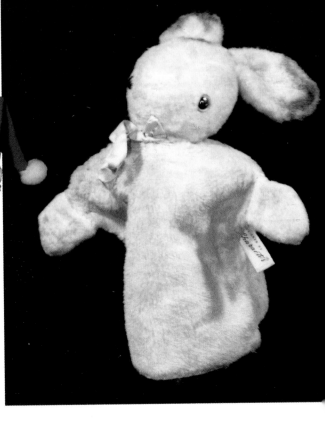

Soft plush apple and wiggling worm hand puppet. 1980s. $7-12

Plush dragon hand puppet. Japan. 1980s. $6-10

Lady hand puppet. Handmade papier mache, 1970s. $10- 15

Cotton hand puppet. Bunny with story book in tummy. 1990. $8-12

Cotton Dinosaur hand puppet. 1990. $8-12

Santa. Printed hand puppet. Glover. 1970s.
$15-20

Cotton mouse in rubber clock, finger
puppet. 1990. $8-12

Hand knit clown hand puppet. $6-10

Ducks, knit hand puppets. $6-8 pair

Cotton stuffed cloth, Snoopy-like puppy
hand puppet. 1990. $6-10

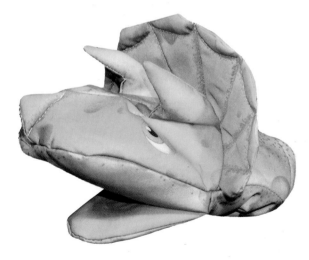

Cloth Dinosaur hand puppet. 1992. $6-10

Rubber

Vinyl head rabbit with cotton dress finger puppet. 1985. $10-15

Rubber Troll hand puppet. 1960s. $10-15

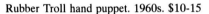

Nurse Troll hand puppet with cotton dress, rubber head, and glass eyes. 1990. $20-30

Troll hand puppet with felt body, rubber head, and glass eyes. 1950s. $35-45

Vinyl head Troll. 1970s. $15-20

Chicky hand puppet with rubber head and
cloth body. 1980s. $10-15

Ninja puppet with vinyl head and gloves.
1980s. $8-12

Rubber head for hand puppet. 1970s.
$10-15.

Princess hand puppet with rubber head.
1970s. $15-20

Santa Claus with vinyl head and cloth body.
Show Character, 1980s. $20-30

Vinyl head farmer. 1960s. $40-50

Rubber head witch from Knickerbocker.
1970s. $15-20

Baby hand puppet with rubber head. $15-20

Queen hand puppet, felt over paper. 1980s. $10-15

Vinyl monkey hand puppet. $6-10

Grandma, vinyl head hand puppet. 1990s. $7-10

Rubber head Devil hand puppet. $15-20

Rubber head hand puppet. European-Gund, 1970s. $15-20

Vinyl king puppet head. 1980s. $8-10

Timbertoes rubber finger puppets from
Highlights magazine. 1970s. $20-30 set

Rubber head Sparky the Firedog hand
puppet. 1970s. $20-25

Rubber shark hand puppet. 1970s. $10-15

Wood, Composition, Plastic

Hand made wooden dancing puppet. 1960s.
$15-20

Hand-carved wood hand puppet. American
Indian. 1970s. $10-15

SURPRISE PUPPET "IT TALKS"

Open The Bag and Discover The
Assortment of Prizes and Games
Then Use as A Hand Puppet

"The Ideal Gift For Children"

Rubber doctor hand puppet. 1970s.
$25-35

Rubber little girl hand puppet. 1970s.
$25-30

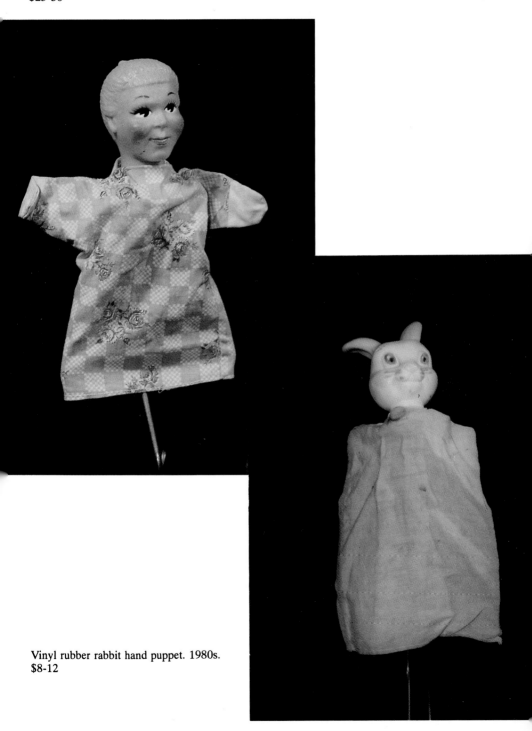

Vinyl rubber rabbit hand puppet. 1980s.
$8-12

Little boy rubber hand puppet. $25-30

Black doctor, hand puppet. Rubber head.
$30-40

Red Robin restaurant hand puppet. Plastic
head. 1980s. $ 15-20

Composition head clown from Germany.
Hand puppet. 1970s. $15-20

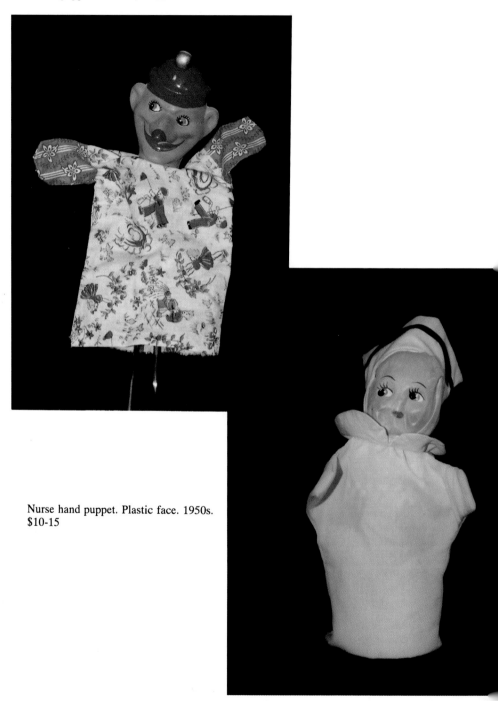

Nurse hand puppet. Plastic face. 1950s.
$10-15

Puppet Books

How to Make Puppets. England. $8-10

Marionettes: A Hobby for Everyone by
Mabel and Les Beaton. $40-45

Dolls and Puppets. Mary Cockett. 1970s. $20-25

Be a Puppet Showman by Remo Bufano. Hard back instruction book. 1950. $30-40

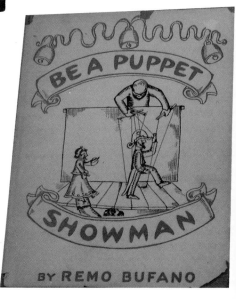

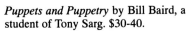

Puppets and Puppetry by Bill Baird, a student of Tony Sarg. $30-40.

Punch and Judy Fight for Pie, Golden Book. 1960s. $25-30

The Tell It and Make It Book by Shari
Lewis. 1980s. $15-20

A Puppet No More. Hard back. 1970.
$10-15

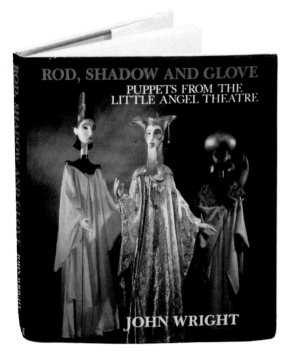

Rod, Shadow and Glove Puppets. John Wright. 1980s. $20-30

Singing-a-long with my Puppet Pals. 1980s. $5-10

Pressed board puppet theater. 1950s.
$100-150

Shari Lewis Flip Winks. 1970s. $10-15

Colorforms from "The Muppet Show" by
Jim Hensen. 1980. $15-20

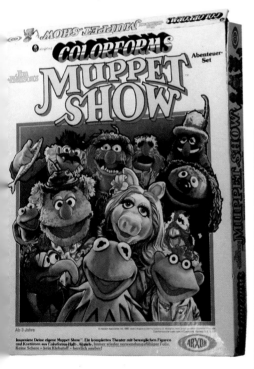

Puppets. Hard back. 1970. $10-15

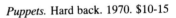

Bibliography

Hockenberry, Dee. The Big Bear Book. Atglen, PA: Schiffer Publishing, 1996.

Joly, Mike. *Toy Box.* Volume 2, Number 2. Troy, MI: 1993.

————. *Toy Box.* Volume 1, Number 1. Troy, MI: 1992.

————. *Toy Box.* Volume 1, Number 3. Troy, MI: 1992.

Mandel, Margaret Fox. *Teddy Bears and Steiff Animals* Paducah, KY: Collectors Books, Schroeder Publishing Co. Inc. 1987

The Puppetry Journal. Vol. IX-no.2., September-October, 1957. Ashville, Ohio.

Toy Trader. Volume 1 Issue 2. Dubuque, Iowa: October, 1993.